STUDENT GUIDE

GETTING IN SHAPE

EXPLORING TWO-DIMENSIONAL FIGURES

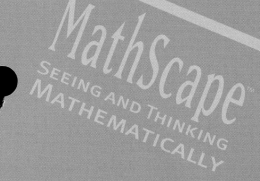

MathScape
SEEING AND THINKING MATHEMATICALLY

What
relationships
exist among
two-dimensional
figures?

GETTING IN
SHAPE

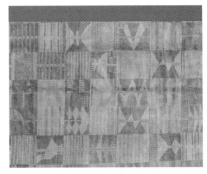

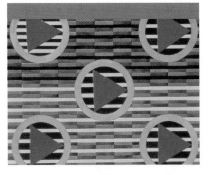

PHASE**ONE**
Triangles

In this phase, you will estimate angle measures and draw and measure different types of angles. You will also investigate various angle and side relationships in triangles, explore the different ways you can classify triangles, check for lines of symmetry, and explore similarity in triangles. Finally, you will enlarge a logo and determine what is involved in creating similar triangles.

PHASE**TWO**
Polygons

You will develop your own classification system for sorting a set of Polygon Tiles™ and explore how different four-sided figures are related. Then you will investigate angle relationships in polygons and discover your own formulas. You will expand your knowledge of congruence to include polygons and explore how to use transformations to determine whether polygons are congruent. You conclude by investigating congruence using transformations on the coordinate plane.

PHASE**THREE**
Circles

You will investigate the unique properties of circles and discover how the circumference and the diameter of a circle are related. The activities that follow will lead you to develop formulas for finding the area of a regular polygon and the area of a circle based on the familiar formula for the area of a rectangle. Finally, you will create a design using figures you have studied in this unit and write a report that explains the mathematical relationships behind your design.

PHASE ONE

In this phase, you will explore triangles by examining angles and sides. You will develop methods for describing and comparing triangles, including how to determine whether triangles are similar or not.

Logos often include geometric shapes such as triangles. Maintaining an exact design when replicating a logo is important. You will apply what you have learned about triangles to enlarge a logo while maintaining the shape and proportions of the original design.

Triangles

WHAT'S THE MATH?

Investigations in this section focus on:

GEOMETRY and MEASUREMENT

- Measuring and classifying angles
- Investigating the angles and sides of triangles
- Classifying triangles
- Finding lines of symmetry
- Identifying similarity in triangles

NUMBER

- Counting lines of symmetry

STATISTICS and PROBABILITY

- Recording angle measures and looking for patterns in your data
- Writing true statements about triangles

MathScape Online
mathscape2.com/self_check_quiz

1 The Angle on Angles

When you study geometry, angles are a good place to start. In this lesson, you begin your exploration of two-dimensional figures by creating an angle-maker and using it along with a protractor to investigate angle measurement and angle classification.

Estimate and Measure Angles

How closely can you estimate angle measures?

Follow the directions to create an angle-maker. For these activities, switch roles with your partner after three turns.

1 See how accurately you and your partner can make an angle.

   a. Name an angle measure between 0° and 360° and record the angle.

   b. Have your partner use the angle-maker to show the angle.

   c. Measure and record your partner's angle.

   d. Use a protractor and a straightedge to draw the angle.

2 Try it the other way. Use the angle-maker to show an angle.

   a. Have your partner estimate and record this measure.

   b. Check your partner's estimate with a protractor. Then draw the angle with a protractor.

How to Create an Angle-Maker

1. On each of three sheets of different-colored construction paper, use a compass to draw a large circle about 9 inches across. Then cut out the circles.

2. Use a straightedge to draw a line from the center of each circle to the edge. Then cut along the line to make a slit in each circle.

3. Fit two of the circles together along the slits as shown.

Investigate Three Angles in a Circle

Use an angle-maker with three colors to investigate how many ways three angles can be combined to fill a circle. Here is one possibility.

How many different ways can three angles be combined to fill a circle?

∠UAV	Acute
∠VAW	Acute
∠WAU	Reflex

Keep a written record of all the combinations you find. For each combination, record your results as follows.

- Draw a picture of how the angle-maker looks. Label the angles using letters of your choice.

- Record the names of the three angles that fill the circle.

- Classify each angle.

How to Name and Classify Angles

Angles are named by the point at their vertex. The angle shown is angle A, or ∠A. Sometimes it's helpful to use three letters to name angles. In this case, the vertex is the letter in the middle. We can call this angle ∠BAC or ∠CAB.

Angles are classified by their measures.

∠DEF is **acute.** It measures between 0° and 90°.

∠JKL is a **right** angle. It measures 90°.

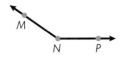

∠MNP is **obtuse.** It measures between 90° and 180°.

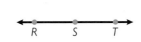

∠RST is a **straight** angle. It measures 180°.

∠XYZ is a **reflex** angle. It measures between 180° and 360°.

hot **words** | angle
vertex

Homework

page 302

2 The Truth About Triangles

Triangles are geometric figures with three angles and three sides. Can any three angles be the angles of a triangle? Can any three sides be put together to form a triangle? Here's a way to investigate these questions.

Explore the Angles of a Triangle

What three angles of a triangle are possible?

Can the three angles of a triangle have any measures, or are there some restrictions? Use the triangular Polygon Tiles and the following ideas to guide your investigation.

1 Trace each triangular Polygon Tile on a sheet of paper.

2 Label the angles and measure them with a protractor. You may need to extend some of the sides so they are long enough to measure.

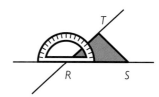

3 Record the angle measures, and then look for patterns in your data. Be sure to consider largest possible angles, smallest possible angles, the sum of the angle measures, and so on.

4 Write down any conclusions you can make about the angles of a triangle.

Explore the Sides of a Triangle

Can any three side lengths be put together to make a triangle? If not, how can you tell whether three given side lengths will form a triangle? You can investigate this using strips of paper as follows.

■ Cut strips from a sheet of notebook paper. Each strip should be as wide as one line on the paper. Cut strips of these lengths: 3 cm, 4 cm, 5 cm, 6 cm, 7 cm, 8 cm, 9 cm, and 10 cm. Label each strip with its length.

What can you say for sure about the sides of a triangle?

■ Choose three strips and place them together to see if they form a triangle.

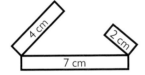 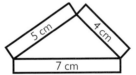

■ As you experiment, keep a record of which strips you used and whether or not they formed a triangle.

■ Write down any conclusions you can make about the sides of a triangle.

hot words | triangle

Homework
page 303

3 Can a Triangle Have Four Sides?

Like angles, triangles can be classified in different ways.
Knowing the different types of triangles makes it easier to talk about them and investigate what is and is not possible.

Explore the Possibilities

What combinations of side and angle classifications are possible for a triangle?

There are three ways to classify triangles by their sides and four ways to classify triangles by their angles.

1. Choose a pair of triangle classifications, such as scalene and obtuse, to investigate. Is it possible to have such a triangle?

2. Use a ruler, a protractor, and/or the triangular Polygon Tiles to find out whether such a triangle is possible. If it is possible, draw an example. If it is not possible, write an explanation.

3. Explore all of the combinations. (Be sure you have them all!) You may want to record your work in a chart.

Types of Triangles

Triangles can be classified by their sides.

Scalene

In a scalene triangle, no sides have the same length.

Isosceles

An isosceles triangle has at least two equal sides.

Equilateral

An equilateral triangle has three equal sides.

Triangles can also be classified by their angles.

Acute

An acute triangle has three acute angles.

Right

A right triangle contains a right angle.

Obtuse

An obtuse triangle contains an obtuse angle.

Equiangular

An equiangular triangle has three equal angles.

Relate Symmetry to Triangles

Is it possible to draw a triangle with the following number of lines of symmetry? If so, draw an example and tell what type of triangle it must be.

1 No lines of symmetry

2 Exactly one line of symmetry

3 Exactly two lines of symmetry

4 Exactly three lines of symmetry

5 Four or more lines of symmetry

What can you say about the lines of symmetry for different types of triangles?

Looking for Symmetry

A figure has *symmetry* (or is *symmetric*) if there is at least one line that divides it into two halves that are mirror images of each other.

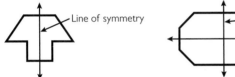 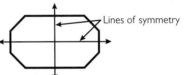

Write True Triangle Statements

Using the language and ideas of this lesson, write as many true statements about triangles as you can. Here are some examples of good statement openers to get you started.

- All isosceles triangles have . . .

- Every scalene triangle . . .

- No right triangle is also . . .

- Every triangle with three lines of symmetry . . .

hot **words** | symmetry
line of symmetry

page 304

Enlarging Triangles

In this lesson, you will learn some terms that are used to compare triangles. Two triangles are called **similar triangles** if one looks like an exact enlargement of the other. How do you decide whether two triangles are similar?

Create an Enlargement

What are some methods for enlarging triangles?

The Student Council at Fort Couch Middle School has been asked to help decorate the Triangle Café by making posters of its logo. Follow the instructions to make an enlargement of the logo using grid paper that your teacher will provide. After you have made the enlargement, answer the questions below.

1 How does your enlarged logo compare to the original?

2 Are the angles the same? Are the lengths of the sides the same? Is it the same type of triangle?

3 The Student Council has requested that you put an enlarged copy of the logo on one wall of the café. Write down as many ways as you can think of to accomplish this task.

Making a Grid Enlargement

Step 1 Carefully examine the logo on the small grid paper and choose one box to use as a starting point.

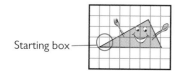

Starting box

Step 2 Find the corresponding box on the larger grid paper.

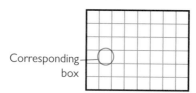

Corresponding box

Step 3 Copy the starting box so that it fits into the larger box as shown.

Step 4 Repeat Steps 1–3 until the entire figure has been enlarged onto the larger grid paper.

Explore Similarity

Using the handouts your teacher will provide, examine the enlarged triangles created by the class. Analyze how each triangle compares to the original triangle. After you have completed the chart, answer the following questions.

1 Which triangle or triangles are exact enlargements of the original logo?

2 What generalizations can you make about similar triangles?

What must be true for two triangles to be called *similar* triangles?

Definitions

Corresponding parts are parts on the enlarged figure that match the original figure. For example, angle A and angle D are corresponding parts. Line segments AB and DE are corresponding parts. Can you name other corresponding parts?

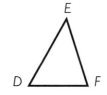

Congruent angles are angles that have exactly the same measure.

Congruent triangles are exact copies of the same triangle.

Letter to Students

Choose a student whose triangle is not similar to the original logo and write a letter to him or her explaining why the triangle does not meet the Student Council's request. In your letter, be sure to discuss the following:

- Explain the meaning of similarity.

- Explain how the triangle does not meet the definition of a similar triangle.

- Be specific about which parts of the triangle are drawn incorrectly.

- Use the appropriate terminology to describe the triangles. For example, specify scalene, isosceles, or equilateral. Also specify acute, right, obtuse, or equiangular.

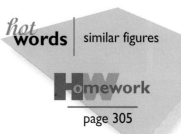

hot **words** | similar figures

Ho**mework**
page 305

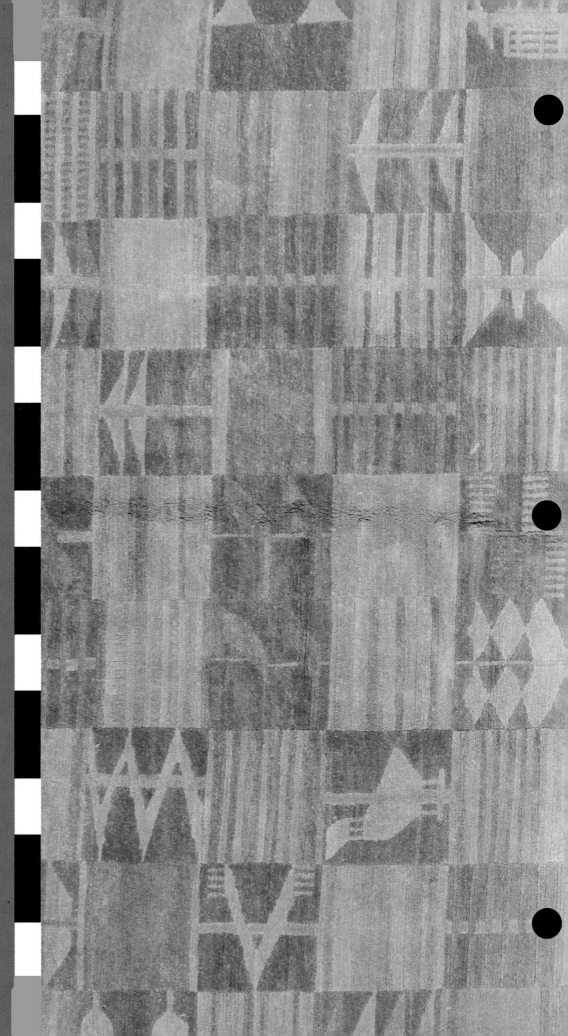

PHASE TWO

In this phase, you will make some fascinating discoveries about polygons. You will find out how different types of four-sided figures are related, explore angle relationships, and look into the relationship between symmetry and regular polygons. Finally, you will explore congruence.

Professionals who prepare blueprints rely on their knowledge of geometry to display in detail the technical plans architects and engineers draw up for their clients.

Polygons

WHAT'S THE MATH?

Investigations in this section focus on:

GEOMETRY and MEASUREMENT

- Exploring different types of quadrilaterals

- Creating tessellations

- Finding lines of symmetry in polygons

- Analyzing the relationship between symmetry and regular polygons

- Exploring congruence using flips, turns, and slides

- Investigating flips (reflections), turns (rotations), and slides (translations) on the coordinate plane

- Writing about transformations

STATISTICS and PROBABILITY

- Developing a classification system for polygons

- Keeping an organized list of results

ALGEBRA and FUNCTIONS

- Writing a formula for angle relationships in polygons

MathScape Online
mathscape2.com/self_check_quiz

5 Polygon Power!

CLASSIFYING POLYGONS AND QUADRILATERALS

A polygon is a closed figure, in a plane, with three or more sides. You already know about triangles—the simplest polygons. It makes sense to ask some of the same questions about other types of polygons. Classifying figures is always a good place to start; then you can begin investigating what is and is not possible.

Sort Polygons

What are some different ways that polygons can be classified?

Sorting polygons helps you develop your own ways of classifying them. Begin with a complete set of Polygon Tiles. Work with classmates to sort them in three different ways that make sense to you.

Be sure to take notes on how you sort the Polygon Tiles, since you may be asked to describe your methods to the class.

Number of Angles

3	4	5	6	7
△ ◿	■	⬠		

Types of Quadrilaterals

Here are some types of quadrilaterals you should know.

Parallelogram

A parallelogram is a quadrilateral with two pairs of parallel sides.

Rectangle

A rectangle is a quadrilateral with four right angles.

Square

A square is a quadrilateral with four right angles and all sides of equal length.

Rhombus

A rhombus is a parallelogram with all sides of equal length.

Trapezoid

A trapezoid is a quadrilateral with exactly one pair of parallel sides.

Isosceles Trapezoid

An isosceles trapezoid is a trapezoid with nonparallel sides of equal length.

Investigate Quadrilateral Relationships

As you explore the following with classmates, keep track of your results using words and/or drawings.

How are the different types of quadrilaterals related?

1 Find all the quadrilateral Polygon Tiles that fit each definition. Which definitions have the most Polygon Tiles? the fewest?

2 Now try it the other way around. For each quadrilateral Polygon Tile, find all the definitions that apply to it. Which Polygon Tiles have the most definitions? the fewest?

3 Write as many statements as you can about the relationships among quadrilaterals. For example, "Every square is also a. . . ."

Summarize Quadrilateral Relationships

Find a way, such as a "family tree," Venn diagram, or some other method, to summarize the quadrilateral relationships you examined. Explain how your summary works.

hot **words** | polygon
quadrilateral

Home**work**

page 306

6 Standing in the Corner

INVESTIGATING
ANGLES OF
POLYGONS

The sum of the angles of any triangle is 180°. Can you say something similar about the sum of the angles of other polygons? First you will investigate this for quadrilaterals; then you will look for patterns that work for any polygon.

Explore Angles of Quadrilaterals

What can you say about the sum of the angles of a quadrilateral?

Use the four-sided Polygon Tiles to find out about the sum of the angles for any quadrilateral.

1 Choose a four-sided Polygon Tile and carefully trace it on a sheet of blank paper.

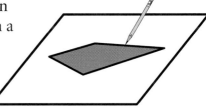

2 Extend the sides of the quadrilateral so you can measure the angles with a protractor.

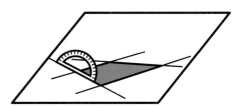

3 Measure the four angles and calculate their sum. Record your findings.

4 Repeat the process with other four-sided Polygon Tiles until you are ready to make a generalization.

Explore the Angles of Any Polygon

What can you say about the sum of the angles of polygons in general?

You know that three-sided polygons (triangles) have angles that add up to 180°, and you just investigated this for polygons with four sides (quadrilaterals). What happens with polygons that have five or more sides?

Each member of your group should choose a number from those shown below. Everyone should have his or her own number. You will be responsible for investigating the sum of the angles of polygons with the number of sides you choose.

	5	6	7	8	9	10

1. Choose a Polygon Tile that has the number of sides you are investigating and carefully trace it on a sheet of blank paper. You can also draw your own polygon.

2. Extend the sides of the polygon so you can measure the angles with a protractor.

3. Measure the angles and calculate their sum. Record your findings.

4. Repeat the process with other polygons that have your chosen number of sides until you are ready to make a generalization about the sum of their angles.

> I chose 6, so I'm doing hexagons.

> The heck with that—I'd better get started on these octagons.

Look for a Pattern

Write a description of any patterns you see in the record you kept. Use words, formulas, and/or equations.

Once you know the number of sides, you should be able to use your results to find the sum of the angles of any polygon.

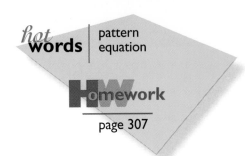

hot **words** | pattern
equation

Homework
page 307

7 Moving Polygons Around

You've already learned about congruent triangles. What do you think it means if two *polygons* are congruent? How can you show that two polygons are congruent? In this lesson you will explore ways to determine whether polygons are congruent.

Explore Congruence

How can you decide whether polygons are congruent?

Look at the cards your teacher has given you and compare them to each of the figures below. Decide whether each figure on the card is congruent to any of the figures below. Be prepared to explain your reasoning.

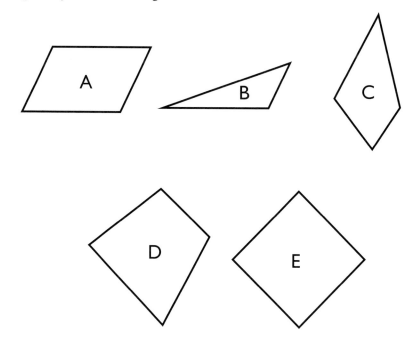

Investigate Polygons on the Coordinate Plane

Use what you know about flips, turns, and slides to answer each question. Refer to the handout provided by your teacher.

1. Write the coordinates for each vertex of figure 1 and figure 2.

2. Describe the move that occurred from figure 1 to figure 2 as a *slide, flip,* or *turn.*

3. Predict the coordinates for each vertex in figure 2 if it is shifted up 2 units. Check your prediction by graphing the figure.

4. Write a general rule for how the coordinates (ordered pairs) of the vertices change for vertical slides.

5. Write the coordinates for each vertex of figure 3.

6. Describe the move that occurred from figure 2 to figure 3.

7. Predict the coordinates for each vertex in figure 3 if it is shifted 4 units to the right. Check your prediction by graphing the figure.

8. Write a general rule for how coordinates (ordered pairs) change for horizontal slides.

9. What move happened from figure 4 to figure 5?

10. What move happened from figure 6 to figure 7?

> **How does the coordinate plane help you determine whether two polygons are congruent?**

Writing Transformations

Write directions telling your partner how to transform figure 4. You must include at least one vertical and one horizontal slide. Graph the transformation on your own before you give the directions to your partner. Exchange directions with your partner and follow each other's directions. Check your partner's work using your graph.

hot **words** | reflection (flip)
rotation (turn)
translation (slide)

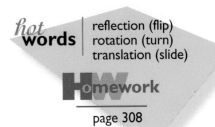

Homework
page 308

8 Symmetric Situations

You have probably noticed that some polygons look nice and even, while others look dented or lopsided. How a polygon looks often has to do with symmetry. In this lesson, you will explore this connection and see how symmetry is related to regular polygons.

Explore Lines of Symmetry

How are lines of symmetry related to the shape of a polygon?

Use Polygon Tiles to explore the lines of symmetry of polygons. Follow these steps.

1 Choose a Polygon Tile and trace it.

2 Use a straightedge to draw all of the polygon's lines of symmetry.

3 Repeat this with other polygons, and keep track of your results. Try a wide variety of polygons, and look for patterns in your results.

What can you say about the lines of symmetry of regular polygons?

Write About Polygons

How are two polygons alike, and how are they different?

Your teacher will provide you with two cards, each of which has a pair of polygons pictured on it. For each card, prepare a report using the directions that follow. Whenever possible, include illustrations.

- Classify each polygon in as many ways as possible using all the terminology you know about polygons. For example, what is the name of the polygon? Is it regular, isosceles, scalene, equilateral, concave, or convex?

- Find the sum of the angles of each polygon. Include anything else you can say about the angles.

- Tell how many lines of symmetry each polygon has.

- Are the polygons similar to each other? Describe why or why not, using corresponding parts of the polygons.

- Is one polygon related to the other by a flip, turn, or slide? Explain in detail.

- Are the polygons congruent? Describe how you know.

Be sure to label each of your two reports with the option letter and card numbers, so it is clear which polygons you are comparing. In addition, staple your cards to each report when you turn them in.

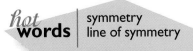

hot **words** | symmetry
line of symmetry

H omework
page 309

PHASE THREE

In this phase, you will come "full circle" in your exploration of two-dimensional figures. After investigating a relationship that makes circles special, you will find out how circles and polygons are related. Then you will create a geometric design based on the mathematical ideas in this unit.

Art museums throughout the world have abstract art collections. Many of these works of art use geometric ideas presented in vivid colors.

Circles

WHAT'S THE MATH?

Investigations in this section focus on:

GEOMETRY and MEASUREMENT

- Measuring the diameter and circumference of circular objects
- Forming rectangles from other polygons
- Investigating the relationship between the area of a rectangle and the area of a circle
- Creating a geometric design

ALGEBRA and FUNCTIONS

- Writing an equation for the relationship between diameter and circumference
- Developing and using formulas to find the area of a polygon and the area of a circle

STATISTICS and PROBABILITY

- Collecting, displaying, and analyzing measurement data

MathScape Online
mathscape2.com/self_check_quiz

 **Going Around in Circles**

Circles are the next stop on your exploration of two-dimensional shapes. You will soon see how circles are related to the polygons you have been working with. For now, you will investigate a fascinating relationship between two of the measurements of any circle.

Measure Circles

How can you measure the diameter and circumference of a circle?

- Your goal is to measure the diameter and circumference of some circular objects as accurately as possible.

- You may use string, a measuring tape, a meterstick, or other tools to help you. All of your measurements should be in centimeters.

- Record your results in a table like the one shown here.

Object	Diameter	Circumference
Jar lid	10.4 cm	32.7 cm

- Be ready to describe your measurement methods to the class.

Parts of a Circle

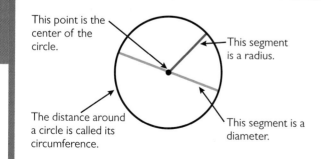

This point is the center of the circle.

This segment is a radius.

The distance around a circle is called its circumference.

This segment is a diameter.

Sometimes *radius* refers to the length of a radius of a circle. The radius of this circle is 1.5 cm. Similarly, *diameter* may refer to the length of a diameter of a circle. The diameter of this circle is 3 cm.

Analyze the Data for Circle Relationships

Use your measurement data to analyze the relationship between the diameter and the circumference of a circle.

What is the relationship between diameter and circumference?

- Describe in words how the diameters and circumferences you measured are related.

- Make a scatterplot on a coordinate grid for your pairs of data points. Describe any relationships you see.

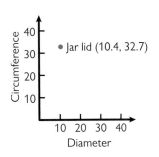

- Describe the relationship between diameter and circumference with an equation. Be as accurate as you can.

Apply Circle Relationships

The radius of Earth is approximately 4,000 miles. Write a brief paragraph outlining what this fact tells you about Earth's measurements, including diameter and circumference. Also provide an explanation of how you found your results.

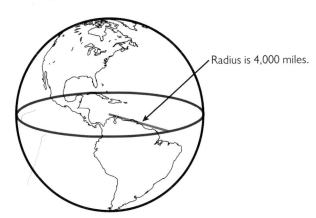

Radius is 4,000 miles.

hot **words** | circumference pi

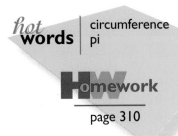
Homework
page 310

10 Rectangles from Polygons

FINDING THE AREA
OF REGULAR
POLYGONS

Before tackling the area of circles, it is helpful to investigate the area of regular polygons. As you will see, regular polygons can be divided up into triangles. Thinking of regular polygons in this way allows you to discover a formula for their area.

Can any regular polygon be rearranged to form a rectangle?

Turn Polygons into Rectangles

One of the special things about regular polygons is that you can divide them into wedges of the same size and shape. To do this, you start at the center and draw a line to each vertex of the regular polygon. If you divide a regular polygon into wedges like this, you can arrange the wedges to form a rectangle. Your goal is to find a way to rearrange any regular polygon so that it forms a rectangle.

1 Choose a Polygon Tile that is a regular hexagon or regular octagon. Trace it on a sheet of paper, and then cut out the polygon.

2 Draw lines from the center of the polygon to each vertex. Cut out the wedges.

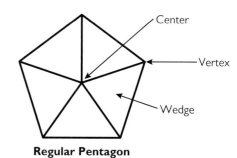

Regular Pentagon

3 Find a way to rearrange the wedges to form a rectangle. You may need to cut one or more of the wedges into smaller pieces.

4 Make a drawing that shows how you used the pieces of the polygon to form a rectangle.

Develop a Formula for the Area of a Regular Polygon

Look back at the rectangle you formed from the pieces of the regular polygon. Then discuss these questions with classmates.

- How is the area of the rectangle related to the area of the regular polygon?

- How can you find the area of the rectangle?

- How does the length of the rectangle relate to the regular polygon?

- How does the width of the rectangle relate to the regular polygon?

- What formula would help you find the area of the regular polygon?

Can you come up with a formula for the area of a regular polygon?

Try Out the Formula

Now it's time to put your formula to the test. Keep a record of your work as you follow these steps.

1 Choose a regular Polygon Tile and measure one side of it.

2 Find the perimeter of the Polygon Tile.

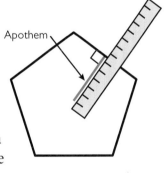
Apothem

3 Locate the center of the tile as accurately as you can. Then measure the apothem of the tile. The **apothem** is the perpendicular distance from the center to a side.

4 Use your measurements and your formula for the area of a regular polygon to find the area of your Polygon Tile. Be sure to include units in your answer.

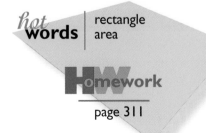
hot **words** | rectangle area

Homework

page 311

11 Around the Area

Now that you have found the area of a regular polygon, you will see that finding the area of a circle is not so different. After estimating the area of a circle, you will develop a formula for the area of a circle using a method similar to the one you used for a regular polygon. Then you will apply the formula to see how good your original estimate was.

Estimate the Area of a Circle

How can you use centimeter grid paper to estimate the area of a circle?

Work with a partner to estimate the area of a circle. Use the following steps.

1 Choose a radius for your circle. Use either 5 cm, 6 cm, 7 cm, 8 cm, or 9 cm.

2 Use a compass to draw a circle with this radius on centimeter grid paper.

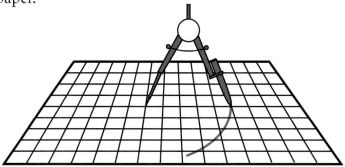

3 Find a way to estimate the area of your circle. Make as accurate an estimate as you can. Keep a written record of your results and be ready to share your method with the class.

Develop a Formula for the Area of a Circle

Can you find a formula for the area of a circle?

Follow the steps for turning a circle into a rectangle. Sketch your results. Then get together with classmates to discuss these questions.

- How is the area of this "rectangle" related to the area of the circle?

- If your "rectangle" were a true rectangle, how could you find its area?

- How does the length of the "rectangle" relate to the circle?

- How does the width of the "rectangle" relate to the circle?

- How can you write a formula for the area of the circle?

How to Turn a Circle into a Rectangle

1. Use a compass to draw a large circle on a sheet of paper. Cut out the circle.

2. Fold the circle in half three times. When you unfold the circle, it will be divided into equal parts or sectors.

3. Cut along the fold lines to separate the sectors.

4. Arrange the sectors so that they form a figure that is as close to a rectangle as possible.

Use the Formula to Evaluate Estimates

Write a brief summary of your findings in this lesson. Include the following:

- the formula for the area of a circle

- the area of the circle you drew at the beginning of the lesson, calculated using the formula

- a comparison of your original estimate and the calculated area

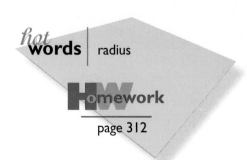

hot words | radius

Homework
page 312

12

Drawing on What You Know

From the repeating patterns that decorate our clothing to the tessellations created with floor tiles, geometric designs are all around us. Behind every geometric design are some mathematical properties that you have explored in this unit. This lesson gives you an opportunity to illustrate these mathematical ideas by creating a geometric design of your own.

Create a Geometric Design

How can you make a geometric design that shows what you know about triangles, polygons, and circles?

It's your turn to create an original geometric design. You may use a ruler, protractor, compass, Polygon Tiles, or any other tools. Follow these guidelines:

- Your design should illustrate at least five mathematical concepts from this unit. Include more mathematical concepts if you can.

- Your design must include triangles, polygons with more than three sides, and at least one large circle.

- You may use color in your design.

Geometric Designs

These photographs are close-ups of a honeycomb and a piece of woven cloth from Africa.

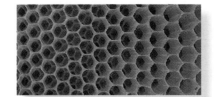
Honeycomb

Cloth pattern

Write a Report on Your Design

Prepare a report to accompany your geometric design. Here are some guidelines on what to include:

- An overall description of your design and the tools you used to create it

- A description of all of the mathematical ideas in your design with additional drawings, if necessary, to explain the mathematics

Be sure to use the correct vocabulary as you describe the geometric ideas in your design.

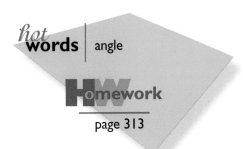

hot**words** | angle

Homework
page 313

The Angle on Angles

Applying Skills

Estimate the measure of each angle.

1.

2.

3.

4.

5. Use your protractor to measure the angles in items **1–4**.

6. Make a table comparing your estimates in items **1–4** with your measurements in item **5**.

Name each angle in three different ways. Classify the angle as acute, right, obtuse, straight, or reflex.

7.

8.

9.

10.

Use a protractor and straightedge to draw an angle with each measure. Label each angle with three letters.

11. ∠KLM, 38°   **12.** ∠CDE, 70°

13. ∠XYZ, 150°   **14.** ∠QRS, 95°

Extending Concepts

There are eight ways to combine three angles to fill a circle. One way is to use two acute angles and a reflex angle as shown.

15. How many ways can three angles be combined to form a straight angle? For each possibility draw a picture and classify the three angles.

16. Suppose that four acute angles are combined to form a straight angle as shown. What is the total number of acute angles formed? obtuse angles? reflex angles? List the names of the angles of each type.

Making Connections

The pie chart shows the breakdown by age of the population of the United States in 1990. The size of each sector is proportional to the percentage in that category.

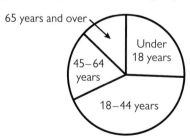

17. For each sector, measure the angle using your protractor. Classify the angle as acute, obtuse, reflex, or right.

18. Calculate the percentage of the population in each age group. Explain how you calculated the percentages.

The Truth About Triangles

Applying Skills

Two of the angle measures of a triangle are given. Find the measure of the third angle.

1. 25°, 80°   **2.** 65°, 72°

3. 105°, 8°   **4.** 12°, 14°

Tell whether each set of side lengths could be used to form a triangle.

5. 7, 9, 10   **6.** 4, 3, 10

7. 2.4, 5.7, 8.0   **8.** 10.8, 13.9, 3.1

Two side lengths of a triangle are given. What can you say about the length of the third side?

9. 8 in., 15 in.   **10.** 21 cm, 47 cm

11. 2.1 m, 2.8 m   **12.** 10 ft, 14 ft

13. Sketch a triangle with angles 59°, 60°, and 61°. Label the angle measures. Put an X on the longest side.

Extending Concepts

14. A triangle has two angles with the same measure.

a. What can you say about the lengths of the sides opposite the two equal angle measures?

b. Could the two angles with the same measure be right angles? obtuse angles? Explain your thinking.

15. The post office is 100 yards from the bank and 50 yards from the store. The bank, store, and post office form a triangle.

a. What can you say about the distance between the store and the bank?

b. The library is equally far from the bank and the post office. If the library, bank, and post office form a triangle, what can you say about the distance between the library and the bank?

c. Which is shorter, walking directly from the post office to the bank or walking first from the post office to the store and from there to the bank? How does this relate to what you have learned in this lesson about the sides of a triangle? Draw a sketch and write an inequality.

Writing

16. Answer the letter to Dr. Math.

Dear Dr. Math:

My friend Lena said that a 179° angle is the largest possible angle in a triangle. But I know that's not right because you can have a $179\frac{1}{2}°$ angle. Then my friend Alex said well how about a $179\frac{3}{4}°$ angle? or a $179\frac{7}{8}°$ angle? Now I'm confused. Can you tell us what the maximum possible angle in a triangle is?

Angus in Angleside

Can a Triangle Have Four Sides?

Applying Skills

Classify each triangle by its sides and by its angles.

1.

2.

3.

4.

Tell whether each statement is true or false.

5. Every obtuse triangle is a scalene triangle.

6. Every isosceles triangle has at least one line of symmetry.

7. No right triangle has three lines of symmetry.

Tell how many lines of symmetry each figure has.

8.

9.

10.

11.

Sketch an example of each type of triangle.

12. right, isosceles

13. scalene, obtuse

14. acute, one line of symmetry

Extending Concepts

15. A triangle can be classified by its angles as acute, right, obtuse, or equiangular.

   a. For each, tell which is possible: no lines of symmetry, one line of symmetry, three lines of symmetry.

   b. Draw an example of each type and show the lines of symmetry.

The *converse* of a statement "If A, then B" is "If B, then A." Tell whether each statement below is true or false. Then write the converse statement and tell whether the converse is true or false.

16. "If a triangle is an equilateral triangle, then it is an acute triangle."

17. "If a triangle is an isosceles triangle, then it is an equilateral triangle."

Making Connections

18. The Ashanti people of western Africa used brass weights to measure gold dust, which was their currency. Each weight represented a local proverb. The picture shows an example of one such weight. Classify the inner triangle by its sides and by its angles. How many lines of symmetry does it have? How many lines of symmetry does the weight as a whole have?

Enlarging Triangles

Applying Skills

Determine whether the following pairs of triangles are congruent. Write *yes* or *no*.

1.

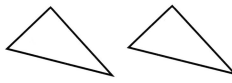

2.

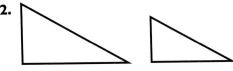

For items 3–6, tell whether each statement is *true* or *false*.

3. Congruent angles have exactly the same measure.

4. Angles are congruent if they are within 5 degrees of one another in measure.

5. A right angle is congruent to a straight angle.

6. An angle that has a measure of 90 degrees is congruent to a right angle.

7. Name each pair of corresponding sides and corresponding angles for the two similar triangles.

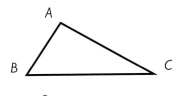

For items 8 and 9, decide whether the following pairs of triangles are similar or not. Describe why or why not.

8.

9.

Extending Concepts

10. Use your protractor for this problem. Are any of these overlapping triangles similar? Explain your answer.

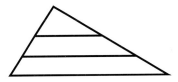

Writing

11. Suppose your friend drew the two triangles below and told you they are similar. Write a note to your friend explaining whether or not you think they are similar. Give at least two reasons to support your answer.

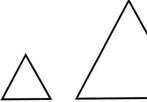

Polygon Power!

Applying Skills

For each polygon, state the number of sides, name the polygon, and tell whether it is convex or concave.

1.    **2.**

Classify each quadrilateral. Give all the names that apply to it.

3.   **4.**

5. Sketch an example of a concave pentagon and a convex heptagon.

Tell whether each statement is true or false.

6. Every rectangle is also a parallelogram.

7. Every parallelogram is also a rhombus.

8. Some trapezoids have two pairs of parallel sides.

For each definition, sketch an example and name the figure.

9. A parallelogram with four right angles

10. A quadrilateral with exactly one pair of parallel sides

Extending Concepts

Tell whether each statement is true or false. If it is false, sketch a counterexample.

11. If a quadrilateral is not a parallelogram, then it is a trapezoid.

12. If a trapezoid has two sides of equal length, it is an isosceles trapezoid.

The *inverse* of a conditional statement is obtained by negating both parts. For example, the inverse of the statement "If a polygon is a pentagon, then it has five sides" is "If a polygon is *not* a pentagon, then it *does not have* five sides."

13. Tell whether the statement "If a quadrilateral is a square, then it is also a rhombus" is true or false. Then write the inverse statement and tell whether it is true or false.

Making Connections

Native American tribes traditionally used beads to decorate clothing, blankets, and bags. The picture shows an example of beadwork created by an Eastern Woodland tribe to decorate a bandolier shoulder bag.

14. Identify as many different polygons as you can. Name each polygon and tell whether it is convex or concave. If you identify any quadrilaterals, tell whether they are one of the special types of quadrilateral.

Standing in the Corner

Applying Skills

Find the sum of the angles of a polygon with:

1. 6 sides **2.** 8 sides **3.** 28 sides

Find the sum of the angles and the measure of each angle of a regular polygon with:

4. 7 sides **5.** 9 sides **6.** 55 sides

Find the measure of each angle of:

7. a regular octagon **8.** a regular pentagon

Find the sum of the angles of each polygon.

9.  **10.**

Extending Concepts

11. The sum of the angles of a polygon with n sides is $S = (n - 2)180°$. How many sides does a polygon have if the sum of its angles is 1,440°?

12. Sketch a convex pentagon. From one vertex draw a diagonal to every other nonadjacent vertex of the pentagon. How many triangles are formed?

13. Repeat the process used in item **12** for a hexagon, a heptagon, and an octagon. How many triangles are formed in each case?

Let n be the number of sides of a polygon and T the number of triangles that are formed when diagonals are drawn from one vertex.

14. Describe in words the relationship between n and T. Write an equation that relates n and T.

15. The sum of the angles of a polygon with n sides is $S = (n - 2)180°$. Use your result from item **14** to explain why this formula makes sense.

Writing

16. Answer the letter to Dr. Math.

Dear Dr. Math:

I used the formula $S = (n - 2)180°$ to find the sum of the angles for this polygon. It has five sides, so I multiplied 180° by 3 to get 540°. To check my answer, I measured the angles and got 90°, 90°, 90°, 45°, and 45°. When I added these angle measures, I got 360°, not 540°. Help! What did I do wrong?

Polly Gone

Moving Polygons Around

Applying Skills

Tell whether each pair of figures is congruent.

1.

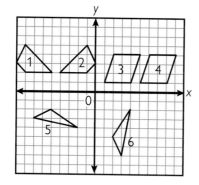

2.

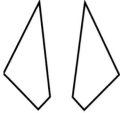

3.

4.

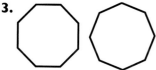

Describe the transformation that has taken place for each figure on the coordinate plane.

5. figure 1 to figure 2

6. figure 3 to figure 4

7. figure 5 to figure 6

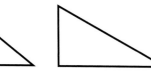

Extending Concepts

For items 8 and 9, refer to figure 4 at the bottom of the first column.

8. Predict the coordinates for each vertex of figure 4 if it were to slide 3 units down.

9. On a piece of graph paper, draw a figure that is congruent to figure 4. Place one vertex at point (3, 3).

Making Connections

Inductive reasoning involves reaching a conclusion by observing what has happened in the past and looking for a pattern. *Deductive reasoning* involves reaching a conclusion by using logic. Each of the following arguments supports the conclusion that the position or orientation of two figures does *not* affect their congruence.

Jose says, "I took 5 polygons on the coordinate plane and transformed them using flips, turns, and slides. None of those changes affected the size or shape of the figure. I measured each one and they were still all the same size and shape. So transformations do not affect whether two figures are congruent."

Margo says, "Since all you are doing when you transform a figure is changing its position — you do not change its size or shape. Therefore, it makes sense that transformations do not affect the congruence of figures."

10. Which type of reasoning did each person use? Which argument seems more convincing to you? Why?

Symmetric Situations

Applying Skills

Tell how many lines of symmetry each polygon has.

1. ☐

2. ▱

3. ⬠

4. ⟨

5. ⌂

6. ⬡

7. ⋈

8. ⯃

9. a regular pentagon

10. a regular polygon with 27 sides

Extending Concepts

11. Make a sketch of each complete polygon by drawing the mirror image of the first half along the line of symmetry.

a.

b.

Then write a report in which you:

• classify each polygon in as many ways as possible and find the sum of the angles

• tell how many lines of symmetry each polygon has

12. Is it possible to draw a pentagon with no lines of symmetry? exactly 1 line of symmetry? exactly 2, 3, 4, or 5 lines of symmetry? Find all the possibilities. Sketch an example of each and show the lines of symmetry.

13. Is it possible to draw a quadrilateral with exactly 2 lines of symmetry? a pentagon with exactly 2 lines of symmetry? a hexagon? a heptagon? In each case, if it is possible, sketch an example. In general, how can you tell whether it is possible to draw a polygon with *n* sides and exactly two lines of symmetry?

Making Connections

It is believed that no two snowflakes are exactly alike. Snow crystals usually have hexagonal patterns, often with very intricate shapes. Each crystal takes one of seven forms depending on the temperature. The pictures below show two of these forms, *plates* and *stellars*.

14. How many lines of symmetry does each crystal have?

Going Around in Circles

Applying Skills

For all problems, use 3.14 for the value of pi. Round answers to one decimal place if necessary.

1. Sketch a circle. Draw and label a diameter, a radius, and the center.

2. If the diameter of a circle is 44 in., what is its radius?

3. If the radius of a circle is 1.9 cm, what is its diameter?

Find the circumference of each circle below.

4. radius = 4

5. diameter = 9

6. radius = 12.8 cm

7. diameter = 18.7 ft

8. [circle with 12 ft diameter]

9. [circle with 5.1 m radius]

Find the diameter and radius of a circle with the given circumference.

10. 3 m

11. 40.7 in.

12. 10.2 cm

13. 0.5 ft

Extending Concepts

14. a. Find the circumference of a circle with a diameter of 8 in. Use 3.14 for π. Is your answer exact or an approximation?

   b. Most calculators give the value of π as 3.141592654. If you used this value of π in part **a**, would your answer be exact? Why or why not?

c. The circumference of a circle with a diameter of 8 in. could be expressed as 8π in., leaving the symbol π in the answer. Why do you think this might sometimes be an advantage?

15. a. Suppose that the radius of a large wheel is three times the radius of a small wheel. How is the circumference of the large wheel related to the circumference of the small wheel? How do you know?

   b. If the radius of the small wheel is 10 in., what is its circumference? How far would it travel along the ground in one complete rotation?

Making Connections

16. a. The diameter of the moon is approximately 2,160 miles. What is its circumference? Give your answer to the nearest hundred miles.

   b. The average distance of the moon from Earth is about 240,000 miles. Assuming that the moon travels in a circular orbit around Earth, about how far does the moon travel in one complete orbit? Give your answer to the nearest ten thousand miles. Explain how you solved this problem.

   [diagram: Moon — 240,000 mi — Earth]

   c. What are some reasons why your answer in part **b** is not exact?

Rectangles from Polygons

Applying Skills

Find the perimeter of each regular polygon.

1. octagon, side length = 5 cm

2. pentagon, side length = 10.5 in.

Find the area of each regular polygon.
Round your answers to the nearest tenth.

3.   **4.**

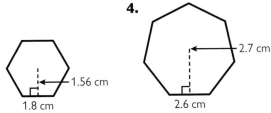

5. regular heptagon, side length = 4.0 ft, apothem = 4.15 ft

6. regular hexagon, side length = 5.4 cm, apothem = 4.68 cm

7. regular octagon, perimeter = 32 in., apothem = 4.83 in.

Extending Concepts

8. Use the regular octagon to answer the questions and draw sketches.

a. Show how the octagon can be divided into same-size triangles.

b. Show how the triangles can be rearranged to make a rectangle.

c. How does the length of the rectangle relate to the octagon? What is its length? How does the width of the rectangle relate to the octagon? What is its width?

d. How does the area of the rectangle relate to the area of the octagon? Why? What is the area of the rectangle? What is the area of the octagon?

9. a. Use the formula $A = l \cdot w$ to find the area of a square with side length 4.

b. What are the apothem and perimeter of a square with side length 4? How did you find them? Now use the formula $A = \frac{1}{2}ap$ to find the area of a square with side length 4. Is this the same as your answer in part **a**?

Writing

10. Answer the letter to Dr. Math.

Dear Dr. Math:

I wanted to find the area of this hexagon. I drew the apothem as I've shown. Then I measured the apothem and perimeter. I used the formula $A = \frac{1}{2}ap$ and got 3.2 cm² for the area, but that seems way too small to me.

I think my teacher got lucky when she used that formula because it only works once in a while.

Do you agree?

K. L. Ooless

Around the Area

Applying Skills

Estimate the area of each circle. Then calculate the area. Use 3.14 for the value of π. Round your answers to one decimal place.

1. **2.** **3.**

4. Tell how the actual areas compare with your estimates.

Find the area of each circle. Use 3.14 for the value of π. Round your answers to one decimal place.

5. radius = 7

6. diameter = 13.5 in.

7. radius = 5 in.

8. radius = 4.1 cm

9.  2.6 cm

10. 1.5 m

Extending Concepts

11. a. A circle with a radius of 10 in. is divided into eight equal sectors. Make a sketch showing how you could rearrange the sectors to make a "rectangle."

b. How does the length of the "rectangle" relate to the circle? What is its length? How does the width of the "rectangle" relate to the circle? What is its width?

c. How does the area of the "rectangle" relate to the area of the circle? Why? If you use $A = l \cdot w$, what do you get for the area of the "rectangle"? What is the area of the circle?

d. If you wanted the figure to look more like a rectangle, what could you do differently when you divide the circle into sectors? Make a sketch to illustrate your answer. Do you think that it makes sense to use the formula $A = l \cdot w$ for the "rectangle" you made even though it is not an exact rectangle? Why or why not?

12. What is the area of the outer circle? of the inner circle? What is the area of the ring? How did you figure it out?

Making Connections

13. A *yurt* is a circular, domed, portable tent used by the nomadic Mongols of Siberia. What would the floor area be in a yurt with a radius of 10 feet?

Drawing on What You Know

Applying Skills

For each polygon used to create the tessellation, give the following information.

1. Name the polygon. For the quadrilateral, give every name that applies.

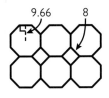

2. Find the sum of the angles and the measure of each angle.

3. Tell how many lines of symmetry the polygon has.

4. Find the perimeter and the area of the polygon to the nearest whole number.

Refer to the design to answer items 5–9.

5. Classify each angle of the triangle.

6. Classify the triangle by its sides and by its angles.

7. How many lines of symmetry does the triangle have?

8. What is the sum of the angle measures of the triangle?

9. Find the diameter, area, and circumference of the circle.

Extending Concepts

10. Create a geometric design of your own. Use a circle, a triangle, and a regular polygon that is not a triangle. Follow the guidelines to write a report on your design.

- Describe your design and how you created it. Is it a tessellation?

- Classify the triangle in as many ways as you can. Classify its angles and tell how many lines of symmetry it has.

- Measure the radius of the circle. Calculate its diameter, area, and circumference.

- Name the regular polygon, and find the sum of its angles. Measure its apothem and side length, and calculate its perimeter and area.

Writing

11. Answer the letter to Dr. Math.

Dear Dr. Math:

We've been learning all these formulas for areas of circles and polygons and for the sum of the angles of a polygon and lots of other things. But who cares about these formulas anyway? Why would I ever need to know the area of a circle? Does anyone ever use this stuff? If so, can you tell me who?

Fed Up in Feltham

Glencoe

This unit of MathScape: Seeing and Thinking Mathematically was developed by the Seeing and Thinking Mathematically project (STM), based at Education Development Center, Inc. (EDC), a non-profit educational research and development organization in Newton, MA. The STM project was supported, in part, by the National Science Foundation Grant No. 9054677. Opinions expressed are those of the authors and not necessarily those of the Foundation.

CREDITS: Unless otherwise indicated below, all photography by Chris Conroy.

271 (tc)©Kumasi Cocoa, photo courtesy of Tufenkian Tibetan Carpets, NY, (tr)Russian State Museum (Liubov Popova), SuperStock; **282** ©Kumasi Cocoa, photo courtesy of Tufenkian Tibetan Carpets, NY; **290** Adam Grand/Getty Images; **292** Russian State Museum (Liubov Popova), SuperStock; **300** Art Wolfe/Getty Images.

Send all inquiries to:
Glencoe/McGraw-Hill
8787 Orion Place
Columbus, OH 43240-4027

ISBN: 0-07-866818-2

4 5 6 7 8 9 10 058 06